The Easy Guide to Podcasting

Published by
The Bourquin Group
1135 Terminal Way #106
Reno, NV 89502
www.bourquingroup.com

Note to readers:

paperback
ISBN # 978-0692320747
ISBN # 0692320741

The Easy Guide To Podcasting

Have you ever wanted to be a DJ, a story teller or a teacher? Have you ever thought about reading books to people?

Are you an author, business owner, consultant or teacher that just wants a better way to reach people?

Podcasting might be your answer.

Writing books can only take your message so far. Some things need to be heard to make sense. Sometimes you can better help your customers, friends or students by talking to them.

Can you imagine trying to call every customer or every person you know every week just to tell them something? With a podcast, you can reach them all on their time. You control the story, they control the delivery.

In this book for example, one of the chapters talks about the differences in microphones. Since sound is about hearing, I wanted you to hear the difference and recorded the same paragraph on three different microphones. That recording is posted as a podcast at www.easyguidebooks.com/podcast so you can hear what the different microphones sound like.

I also added filters so you can hear that too. Reading my opinions in a book doesn't help you form your own opinion or record the best way for you. Only hearing the differences in the microphones can you decide which one is right for you.

The other benefit to podcasting is that people can listen while they drive or ride on the train giving you another opportunity to reach them.

When I set out to create a podcast, I had so many questions and found dozens of answers, some that weren't correct and a few that didn't make sense, and worse a couple of people gave half the information.

I wrote this and recorded the accompanying podcasts so that you can get podcasting as quickly and inexpensively as possible and get your message out there sounding like a pro.

Spread your word!

1. So What Is Podcasting

What is Podcasting? Do you think you know? I thought I knew and tried to record a podcast. It sounded like garbage so I stopped. And then I started digging, and hopefully this book is all the digging you have to do to get started.

I quickly found out that I didn't know what I needed to know, so if you really do know, you can skip this chapter. If you don't know read on....

Since 1998 I have been in the internet marketing business sometimes called "Online Marketing" and followed every trend, trick and technique to keep my clients on top of search results. One day, Google makes a small change and suddenly up pops "podcasting" in the formula. The funny thing is podcasting didn't become really important until after YouTube.

At first glance Podcasting is creating an audio program just like creating a song. You record something and put in on iTunes and people download it to listen.

It turns out Podcasting is very different. To begin with in just about every other method of getting your information out there, you can do it completely free and even get paid. Podcasting is more like radio.

First, if you blog, you can get a Wordpress or blogger account for free. If you want to create a video, there are dozens of sites that will let you edit it for free like Loopster and host it for free like YouTube.

Podcasting isn't that easy. Online recording and editing that works is nearly non-existent. While you might find it out there, I haven't found any that I would say you could use and sound like a pro.

Podcasting requires that you host your own files and then get them approved by Apple for display on iTunes. If you record a song, you upload it to iTunes and sell it. Since you can't charge for a podcast, Apple won't store it for you. You have to pay for that too.

Starting out you can get a low cost web hosting account and upload your first few podcasts.

When you get popular, you may need to pay for a service like Amazon AWS. AWS is very inexpensive in the long run, but you need a plan for making money with your podcast.

Unlike an online video, you can't put Google Adsense ads on the side and hope to make money. So why is podcasting so popular and why is it so important?

Think of it this way, podcasting is your own private radio station that anyone can listen too at any time. But you have to pay for it. If you can get advertisers to place commercials in your podcast, it might cover your expenses.

Most podcasters though have another plan in mind. The podcast is the link to other services. We started looking into podcasting for our clients and realized it would work for us too.

So, what is podcasting? Podcasting is simply recording your own show about something or reading a book or creating any audio content, hosting it and giving it away for free.

So if you have to give it away for free, why do it?

There are several reasons. As the owner of an SEO or Search Engine Optimization company, we look at a link from Apple's iTunes as a "Premium" or "Preferred" link. Links like this are of high value to increasing a companies position on the big three search engines.

If you have a product to sell, a business that wants to reach more clients or if you are in the entertainment industry and want more bookings, Podcasting might be just the tool you want to add to your toolbox.

Podcasting is unique because it is free, and it brings the power and reach of Apple to you or your business for *free*. So what is the catch? The catch is that Apple listens first to make sure you have quality and acceptable content.

So why doesn't everyone podcast? Simple. First, it means you have to sit down and record something. Second, in order to get any kind of traction, you need to have quality content that people want to listen too. If no one listens or you get a few bad reviews, it can be worse than posting nothing at all.

Should you podcast? The question is "Do you want to grow your business or get more bookings?" If the answer is "Yes", then you should podcast.

The Easy Guide to Podcasting will give you all the tools you need to start podcasting and podcast so that people will listen. Steven Covey said that the eight habit of highly successful people was "finding their voice". Podcasting is where you show the world your voice.

Not sure what your "voice" is yet? We can help. Before you call, read the rest of The Easy Guide to Podcasting, and record something, anything even if it is just reading a paragraph from this book.

__Pro Tip -__

No online marketing program is complete without podcasts and video's.

2. The Gear

If you have a ton of money, you can hire a writer, a voice over actor, rent a studio and record a podcast. If you don't, you can spend a little bit of money and set up your own home studio and have fun telling your story, your way.

For nearly 15 years, I have studied acoustics and built sound rooms, home theaters and studios. I have also been able to learn from some great engineers and directors while I recorded in them. A really nice studio can run over a hundred grand to build, and renting one can cost several hundred dollars an hour. That alone would stop most people from podcasting.

There is another way.

The Secret?

For under $1000 you can record studio grade podcasts. For under $300, you can put together a system in your home or apartment that will create very good podcasts. The trick is buying the right gear, and using it right. Keep reading and you'll learn everything you need to know to create your own podcast and not break the bank.

So, how can you make a few hundred sound like a hundred grand? Well some is simple tricks, the rest is understanding sound and your voice. The rest is editing.

For instance a Microphone screen can run up to $300 at the local music store. You can also buy a $20 piece of acoustic foam and $10 tube of glue and glue it to an old piece of cardboard and get the same effect. In your home it is about the sound, not about the aesthetics. If you want it to be pretty, plan to spend more.

When you make the first million and want to invite people in your studio, then you can build the custom room with the $1000 microphones, $10,000 walls and crazy gear. For now, let's keep is simple and easy.

What you really need is a microphone, a computer with software to record and edit, and some place to record. Once you have a recording, you need a place to store it and share it. Most web hosting companies will include about a gigabyte of storage at no charge so your first year or two of podcasts would run about $30 a month on 1and1.com, godaddy.com or several other web hosting companies.

There are links at EasyGuideBooks.com to help you find the sources we have tested and know work.

Microphones, Is There a Difference?

Microphones are all a little different. For the most part you will sound the same on all of them. If you have a very unique voice or a really good ear, you might hear the difference. I can, but I also know some tricks I'll share later.

So you could hear the difference I made a podcast "Which microphone is best". so you can hear what three different microphones sound like to other people.
Like I said before, not everything works in a book.

One of the things I did not test is the internal microphone on the iPhone and iPad. I know from experience that these will not create a quality product for you. I have a friend who webcasts from his car, and says it is with his iPhone. It isn't true, he has a really nice GoPro Camera and a shotgun mike rigged up so the sound is clear in his ragtop cruising to his office.

Basic Types of Microphones.

Omnidirectional - Normally called Omni mic's, Omnidirectional mics pic up sound about equally from any direction. They usually have a very high sensitivity, so they pick up everything in the room, and have a bit of a "tube" sound. The very best models are "boundary" mics, that are designed to sit right on a desk and have multiple pickups and electronics to eliminate echo. These are great for conference rooms, and Facetime, not so great for Podcasting.

Directional - Nicknamed "Shotgun" mics, they are very directional. If you want to record a guitar and a singer with different mics, and not have the guitar mic "hear" the singer, you want a shotgun. News crews on the street use them to "target" the person speaking and not listen to all the street noise. Sports shows with multiple hosts will use shotgun microphones.

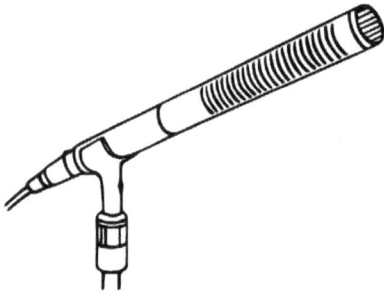

Unless you want to have someone standing around you while you record your podcast so they can keep the mic pointed at you, shotguns shouldn't be your first choice.

Cardioid - Cardioid mics, are typical in studio or on stage mics. The pic up better from the front and a little from all around. This patterns lets you move a little while you talk without making huge changes in how you sound. All of the mic's I would suggest for podcasting are Cardioid

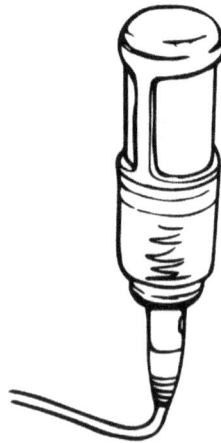

mics. The pattern is a result of two different membranes and electronics that combine the two patterns.

Microphone Connections

Microphones come with one three types of connectors, Analog,XLR and USB.

Analog microphones simply plug into your headset jack or a microphone jack on your PC, Mac phone or tablet. The versions for Blackberry have even smaller connectors. Analog microphones are very inexpensive and usually have very low quality pickups.

Headsets for gaming are an example of an Analog/PC microphone setup. There are some that look really cool and can get very expensive but the reality is they just can't compete with some of the other microphones when it comes to sound quality, depth and clarity.

One of the great secrets of sound discovered by Alexander Bell when he invented the telephone is that you only need to be able to send and receive a very narrow range of sound in order to be understood. If you have you ever known anyone that sounded way different on the phone, then you have experienced this phenomenon.

On the phone my brother and I sound exactly alike. In person we don't. Those nuances in your voice are lost with those low cost analog microphones plugged into your PC. The benefit of these microphones is lower bandwidth usage for using VoIP, FaceTime or internet phone calls. For most computer calls, the tradeoff is worth it. In podcasting it isn't.

XLR microphones are also analog, with a much larger connector. XLR microphones are what the studios use. XLR microphones can be Omnidirectional, Cardiod or Directional. XLR microphones range in price from $29 for budget microphones that look like they belong on a Karaoke machine to $4000 designer studio microphones for people with rockstar wallets or Opera ranges in their voice.

XLR is the type of connector. Most pro studio and stage gear use XLR connectors. XLR is pretty much the studio standard. This way a microphone can be plugged in right next to a guitar, bass, piano or drum set.

The problem with XLR at home is that you need some kind of mixing board in order to record, and more specifically a mixing board that your Mac or PC can connect too. Adding a decent mixing board with USB is the best way to get quality recordings, but this is about doing it the *Easy Way* which doesn't mean the break the bank way. A good one microphone converter like the Apogee One can run over $500.

USB microphones are the lowest cost way to get started in podcasting and still be able to produce a quality product. The problems with trying to classify microphones in the USB category is that you can use USB for everything from webcams, to gaming headsets to dictation headsets as well as near studio grade microphones.

To begin with a microphone should look like a microphone in a studio. When you see it, it should look like it should have an XLR cable sticking out of it. That will be your first clue your are looking at the right microphone for podcasting. Shopping at a music store or pro audio shop would be the next step to making sure you are looking at a good mic. I have seen several "podcasting kits" online, but be careful. Some of those kits have some pretty low quality microphones in them.

There are several very good USB microphones on the market. Blue, makes the Snowball and the Yeti which are great for online phone calls and video calls. I started with the Blue Snowball. After a couple of months and listening to other podcasters, I realized the Blue didn't work for me. I upgraded to the Blue Yeti. Better but still not the sound I was looking for.

On a set with another actor who was there to do a voice over part, told me about his home setup. He had an Apogee MiC and had recorded the audition for that days work on it with his iPhone in a closet. So there is the minimum system requirements. An iPhone, and Apogee MiC and an app to record from the MiC to MP3 that lets you email the file. Needless to say I ordered one before I left the studio. The Apogee MiC was the sound I was looking for. It turned out to be a great purchase. It was so good I only had it for three weeks and then my wife took it.

My wife's first voice coach suggested the AT2020USB because it has a built in headset jack. I bought one hoping that I could trade my wife and get my Apogee MiC back. It turns out there are two models, the AT2020USB and the AT2020USB Plus. The Plus model has the built in headset jack. Of course I bought the wrong one, so it is still sitting on my desk and the Apogee MiC is still on her desk.

This isn't to say that Samson and other brands don't have good microphones for podcasting, these are just the microphones I have personally used and recorded with. Either will give you excellent quality. For the latest updates visit www.easyguidebooks.com and look for podcasts related to *The Easy Guide to Podcasting*.

The better USB microphones have components similar to XLR microphones and small versions of a mixing board that convert the signal to USB. The AT2020 by Audio Tecnica is unique because you can buy an XLR version as well. Ordering this one online requires some careful reading to make sure you don't get the XLR version by mistake.

The Apogee MiC is unique because it can connect directly to the digital port on an iPhone 4, 5 or iPad. This direct digital connection makes for better sound quality and eliminates the need for a separate Mac or PC. It also allows for quick on the fly editing and a very portable recording system.

USB microphones and webcams should never be plugged into a USB hub. Delays and errors can be added by the hubs.

Speakers

After you record yourself, you need to be able to listen and hear every flaw. Some flaws are ok to leave in, keeping your podcast real or natural. Others need to be cleaned up. People listening with noise canceling headsets will hear every flaw. I like having both speakers and headsets on my desk. I use the speakers for the rough edits and the headsets to fine tune anything for the final edit.

My desk has studio grade powered monitors for listening. I have Mackie speakers, and there are several good brands to choose from. I recommend going to a Pro-Audio store and listening for yourself. Unlike stereo speakers, the studio monitors aren't supposed to "color" the sound. Bose will make you sound very different. Studio monitors and studio headsets have a very flat response by design. Some people won't like using them for music listening.

Even with these great speakers on my desk, I still use very nice Etymotic or Shure earbuds to listen to every detail. Yes, I have both, and like them both.

My wife doesn't like ear buds so she has Audio Tecnica studio quality headphones for listening to her recordings before they go to my desk for final clean up. I suggest something like the AT 20 or better. Don't get any of the music headphones like Beats™ or noise canceling like Bose™. They sound great for music but will change what you are hearing.

Studio headphones have a very flat response so you only hear what you record exactly how you recorded it.

If you don't want to clean up and edit your own files, you should start a relationship with a good engineer who will do it for you at a reasonable price.

The Whole Deal

The mini package is simply an Apogee MiC or AT2020Plus, an iPhone or iPad, Studio headphones and budget software like Garage Band or Twisted Wave. For under $700 you have all the gear you need and that includes your phone or iPad.

The next step up is to use a Mac for easier editing, and one of any number of studio grade USB microphones like the Apogee MiC, Audio Tecnica AT2020USB or the Sampson USB and software like Twisted Wave, Garage Band or Audacity.

The mini studio package would use an XLR microphone, an XLR to USB mixer like the Apogee One and a Mac or PC with the software of your choice. This starts pushing $1500 without even talking about the software. ProTools can run over $500 and we haven't even considered acoustic treatment.

Pro Tip -

Most local music stores and guitar stores have someone who knows all about this stuff. See if they have a demo setup you can try before you buy. You never know what you might like. Just don't get talked into buying the high dollar stuff. You aren't recording a grammy winning song.

3. The Room

The second most expensive part of any good sound studio is the room. Most sound studios have two problems. The recording room and the listening room. In the Pro Audio world, each room is equally important.

The listening room is designed to prevent echoes and reflections that can change the sound. These rooms can exceed a hundred grand. You can get everything you need in a reasonably quiet room in your home and headphones that cost less than $100.

The recording room is what you want to work on. This is where you are going to be creating your podcast, and you can only clean up so much stuff. As I mentioned before, you can hear a noticeable hum in my office when the refrigerator in the kitchen is running. I have learned to hear it and guess the timing so I can work around it.

Recently I was working in a pro studio, and in the smaller voice booth there was a big complaint from other voice over actors about being able to hear the engineer from the other booth. They were looking at all kinds of fancy solutions. When my wife stepped in, I just pushed on the door and the problem stopped. The door was slightly out of square and a 5 inch section of the door seal wasn't making contact. A very expensive door was no better than a solid door in your house for that paper thin gap.

When you first start setting up your home or office recording space, you may want to just record a few minutes of "silence" and see what you hear through your headphones. Airplanes, cars, kids, heaters, air conditioners and even computer fans can be quite loud. Another reason I like Mac's for recording. My MacBook Air has no fan and my Mac Mini fan is nearly undetectable.

If you are fortunate enough to have a walk in closet with clothes all around, you might have a great recording room. The problem is getting your computer in there with all the gear. This is where the iPad set up can be really cool. You record to the iPad, move it to your Mac and then edit it there.

Since the iPad doesn't have a fan, it makes a great recording tool. Many of the recording and editing platforms have a version you can put on the iPad and share the files. If not, just get an app that will email .wav and .mp3 files so you can edit them on your computer.

You can edit on a iPad, but the keyboard is infinitely faster and easier. Time is something you can't get back and this is *The Easy Guide to Podcasting*, so I am not going to suggest that you listen and edit on an iPad.

The worst possible room is an empty bedroom with wood or tile floors and mirrored closet doors. Carpeting, beds, chairs pillows all help eliminate the echo that gives that "tube" effect. The empty room is just an echo chamber where the microphone will pick up everything.

There are several websites that will show you how to build really cool simple recording booths in your home. You can use specially made moving blankets, acoustic foam and a number of other tricks to build a little box at home. Personally these boxes get hot and can be claustrophobic for some people.

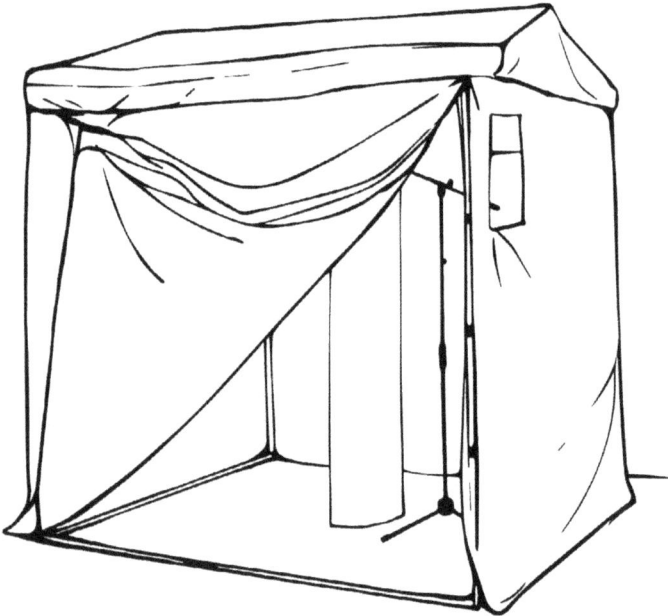

Blanket Room Voice Recording Booth

Any regular bedroom or office will usually work quite well with some simple tools and planning.

Also online are some videos to build a "mic box". Mic boxes work well but make it hard to read your copy while speaking into the box. Sound Shields or Mic Shields can do a big part of that job and can be purchased online or in any Pro-Audio shop.

Basically a shield is a piece of acoustic foam designed to sit behind the mic to stop some echo and other room noise. Generally speaking, the larger the better. The shield in my wives office is quite small because the room is reasonably quiet. My office required a much larger shield.

If you have a bedroom with those big mirrored sliding closet doors, open one side all the way to record. Ideally the open side should be filled with hanging clothes and the closed side should have a piece of soft furniture in front of it like a big overstuffed chair.

Listening carefully to your chosen recording room you should be able to hear everything that makes noise. You will be surprised what makes noise when you start paying attention.

To quickly analyze any room simply stand in the middle and clap. If there is any echo, that is bad. Look at the floor, is there carpet with pad underneath? That is good, wood or tile floors are bad. Along the walls are there bookshelves with lots of irregularly shaped books and objects? That is good. If the walls a flat and just have pictures with glass, that is bad. The ceiling is the toughest because most homes have flat ceilings and that "acoustic ceiling" that was poplar in the 60's and 70's doesn't do enough.

Auralex makes a small pre made foam booth you can fold up and store easily. It isn't exactly what I would call a low budget item. There are other ways to skin the cat.

When you stand in a room, microphone placement is important. You don't want to face a corner or a wall straight on. Nor do you want to be closer than about 4 feet to the wall.

The next step is to simply look at the room. What hard and flat surfaces can reflect sound back to the mic? Floors, Walls and Ceilings all should be considered. If you want to get really serious, you can buy acoustic panels and put them on your walls and ceiling.

If you want to keep your costs down, building a simple foam box around the mic can really make a difference. Mic Shields work pretty good, but they only stop the reflections from the wall behind the mic. They dont' help with ceiling, side wall or floor reflections.

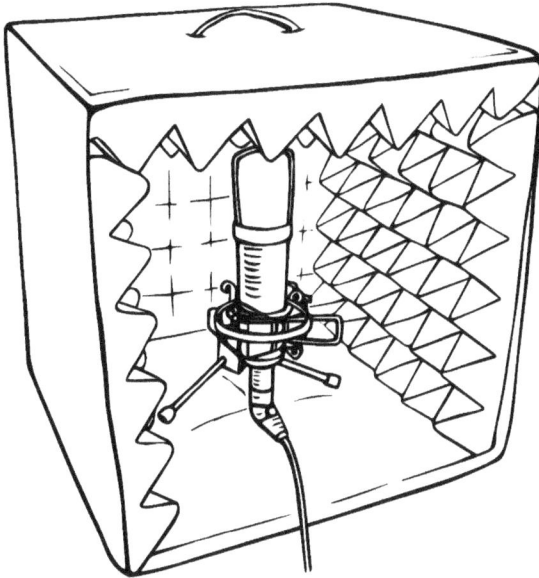

Microphone In A Acoustic Box

The only issue I have heard from people who use the "Mic in a box" is that reading while keeping your mouth aimed at the mic takes some practice.

This doesn't have to be a fancy box. This can literally be an old cardboard box with the acoustic foam cut to fit. You don't even have to waste the glue on the box if you cut the foam just a bit large, it will hold itself in place.

Pro Tip -

Not sure if a room is good? Just sit in it for five or ten minutes. What do you hear. If almost nothing then good. If you hear a fan, a refrigerator or any other constant noise, try another room.

The next step is to clap. Is there an echo? That is bad. Find the room or closet with the least echo after you clap.

4. The Software

To PC or Mac?

To begin with, I prefer Mac's over PC's, however I will say I am not a fan of Apple's Garage Band™ on the Mac. My philosophy which was ingrained in me in the early days of my career while working at Apple was simple. Anyone should be able to use the product without instruction. Apple even created the "Human User Interface Guide" which has evolved into the iOS Human User Interface Guidelines.

My personal experience is that Garage Band™ adds effects as a default, and for podcasting you want your voice, clean and simple. That said, Garage Band™ on the iPad is easier to use and create files. Moving them to Mac for larger edits isn't bad, but Garage Band™ is just different enough from what the studios use that you may want to try something else.

Software is a very personal choice, because it really only matters if you can use it and create the files you need. If you are sending your files out to be edited, you just need a format that the editor can clean up so you sound great.

After taking several classes over the years using Twisted Wave™, Audacity™, CuBase™ and ProTools™, it became clear that ProTools is the industry standard for recording studios, and Cubase is probably the best software out there. For podcasting, both of these are way too expensive and complicated. You aren't mixing 32 channels of music, you just want your voice clean and pure.

Looking at the Apple Ap Store there are over 20 different voice recording Aps, Search online and find even more freeware programs. You do want something that looks and works like studio software so if you ever rent a studio, you will know what is going on and what the engineer is talking about.

Audacity is free for your use but in order to save it in the right format to use as a podcast you have to download other software and and do more work. I got it to work but it wasn't as simple as the online instructions.

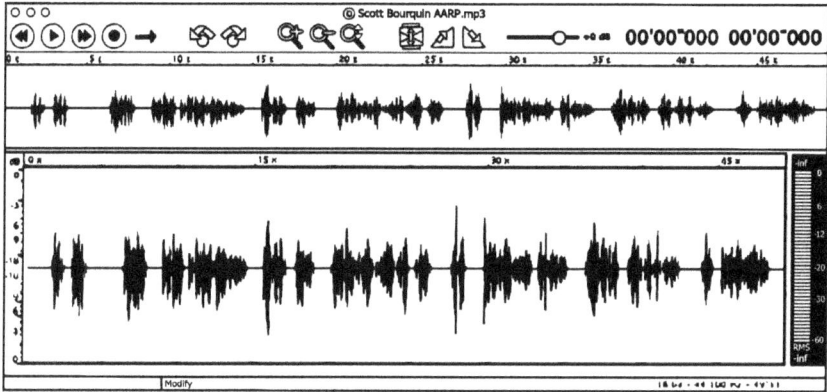

TwistedWave is among the more popular programs for studio voice work and home voice work. The lite version is under $20 at the Apple Ap Store, the full version under $60. A couple of Benjamin's or so less expensive than ProTools.

TwistedWave can record one or two channels and has all of the editing tools needed to clean up and fix just about anything that you need at home for podcasting or even creating webinars with screen capture software like Camtasia.

Any of these recording software packages will let you see your voice, and you can even see the noise in the room, your breathing, lip smacking and in my case I can see when my refrigerator is running.

There are some simple tricks to record silence and paste over all of these flaws so your podcast sounds professional any time. There is an option to "paste silence" which if used right sounds ok, but if there is any flaw in your room that the mic pics up, the "silence" might sound like your podcast ended.

Over time you will hear more and more in your recording room. My office never bothered me until I started recording. I have a large single pane window so I get all the street noise. My wife has double pane windows so her office is quieter. My office shares a wall with the kitchen and the furnace. Both are so loud you can't record, and I never heard them before.

The worst offender is my server. The little network storage device I bragged about in my blog posts is wicked loud when you start listening. I was going to move it to my audio rack, but then I would hear it when I watch TV.

I didn't hear any of this until I started recording and listening.

Pro Tip -

Most Music Stores have a free class you can take to learn about their software. Also there are lots of great online videos to teach you what to look for in all of those wavy lines when you start editing.

Editing can make or break a good podcast.

5. How to set it up

The set up is pretty straight forward in most cases. The first thing is to decide how you are going to record and work outward from there.

We have three different setups in our house because of the different noises in different rooms. For short podcasts, under 3 minutes, my office will work. I just have to work around all of the noise that the mic pics up which gets tedious after a while.

The Closet

I do have a reasonably large walk in closet, so I have made the Apogee MiC in my wives office somewhat portable. Any USB microphone on a stand will do but the Apogee has the second 30 pin cable so it can connect directly to an iPad or iPhone.

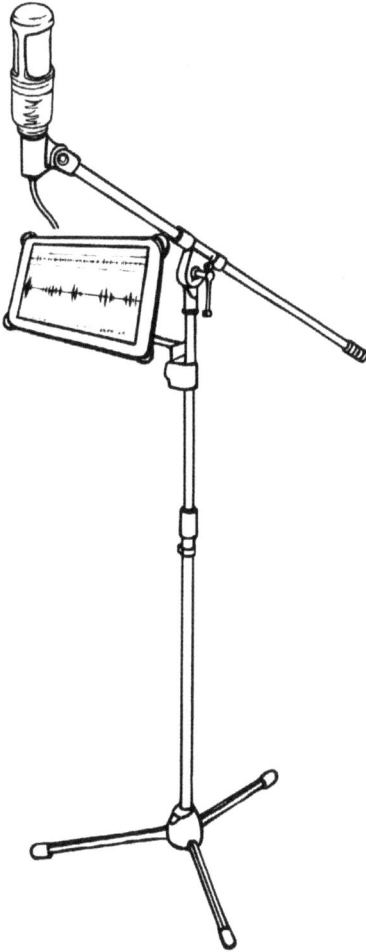

The mic stand remains folded up in the closet until I need it. I can simply unfold it, plug in my MacBook Air or iPad.

The coolest part with this setup is that you can get an iPad mount and have the iPad act like a teleprompter. No paper rustling around and no page turning while reading.

The whole package including iPad $700

The hard part with this little gem is when you mess up, you can't see the timing on the screen like you can with a computer because you are using your iPad as a teleprompter. Also the iPad is a little tough to edit on because of the lack of a real keyboard and the small screen.

Pro Tricks For Pro Sound

The pros in the studio have some great tricks I want to share with you though since starting out with a limited budget is most likely what you need to do. Sitting in the studio is an expensive proposition so we can learn some great tips there. Even the pros getting paid to do all of those cartoon voices, know that every error is money.

The first time I sat in the studio watching a couple of cartoon characters come to life, I expected that if the voice actor made a mistake, the director would say "cut", tell them what they did wrong and then record again.

That does happen but not very often. More often the voiceover artist will simply pause and re-

record until they get it correct. A simple gesture lets the engineer know to keep recording. In the world of digital recording, the is the least expensive way to do it.

At home you can do the same thing. To make it easier to find those places you made a mistake, simply snap your fingers to the mic. You will see a massive jump on the screen, and will know to look there.

If you are getting into longer podcasts or audio books, this technique will save you more time than the cost of this book alone.

Office One

My wife's' office is a simple setup. A Mac Mini, Twisted Wave Lite, two very nice computer speakers with a headphone jack on the front, headphones of course and the Apogee MiC on a stand with a pop filter and Microphone shield.

The whole setup was under $1000.

The Office One Setup

Office Two

My office looks like the macdaddy of home sound gear, but since it doesn't have the acoustic treatments to quiet all the noises that leak in, the recording quality isn't any better.

The setup is a Mac Mini on steroids, Dual 27" screens, a trackball, studio grade monitors, a microphone on a stand with a pop filter. By the time you read this a microphone shield or a shotgun mic will be in place so I can stop running to the closet to record. I might break down and add some acoustic treatment too.

The dual 27" screens make editing much easier. So does the trackball. Originally I had a wireless laser mouse, but I noticed at both the major studios I have worked, all of the engineers stations had trackballs. Once I tried it I was hooked. A trackball is much faster and more accurate than a mouse. Get a trackball.

Potential upgrades would be a mixer panel to go to XLR mic's and maybe even a dual shotgun setup so I could record interviews. That is about it.

As of now, the closet is the best recording location in my house simply because it has the most sound insulation and has bedrooms on three sides so there is no outside noise during the day.

My office (Office Two) is the best editing station.

My wife's office (office one) is the best balance for getting it all done in one place.

Pro Tip -

It is easier to edit tracks with low background noise. If airplanes fly over, do it again, a truck drives by, do it again.

6. Write or Talk?

One of the secrets of the business world is that everything has a script. If you see it on TV or it is professionally produced it has a script. If you want to sound like a pro, then you need to work like a pro and *create a script*. The best speakers and storytellers know the script cold so you can't tell it is a script or worse, a teleprompter.

This doesn't mean you have to read everything from a teleprompter, but you should have it written down and rehearse it before you record it. The best part about recording is that you can read just one paragraph at time until is sounds right an then move on. You don't have to live with a mistake like some live news people.

The big trick is to make your script sound natural and like you are talking to your audience, not just talking to yourself. A technique that was given to me some time ago is to write to a very specific person. Cut a photo out of a magazine. Name the person, give them a back story. Explain how they connect to the work you are doing, and then start writing to them.

You will be amazed how much different it sounds.

When you record, do the same thing. Talk to the picture, what ever you are talking about or podcasting, sell it to them, really sell it.

The other thing about writing it down, is you run it through different parts of your brain. More ideas will be flushed out and you will come up with better content.

Comedians look like they are on stage telling stories for the first time every time. If you even catch the Chris Rock HBO special, you will see the true story. Right in the middle of a joke or story, they will switch to another venue. Chris has rehearsed his show so well, that he gives a nearly identical performance in every city.

On the improv circles Chris Rock famous for dropping into a small club and trying out new material, writing, rewriting and doing it again until the audience is on the floor. Only then does it make his full show.

Think about a Chris Rock show this way, no matter what you think of his comedy, the man gets on stage for over an hour and puts on a one man show, never missing a beat, a line or an expression. In comedy, timing is the show. That is like memorizing hamlet, and doing the entire show on your own. I only have one word for this level of work - impressive.

The short version of the chapter is always write down what you want to say, and edit it at least twice before recording it. Your podcasts will sound more professional and polished. When you get enough practice and that big speaking gig comes in, then go for the cue cards, or be Chris Rock and just put on a stage show that is just all you, polished to perfection.

Pro Tip -

If you are not sure what to write or talk about, go look at your competition. What questions are on the FAQ page that you could answer better? There is a podcast. What questions aren't answered? There is a podcast.

The other thing about writing out your podcasts first, you have an instant blog entry! Remember the idea is to repurpose, and reuse as much of the media you create. If you write something down and don't posted, who will ever know?

7. Using a Mic Correctly

There are two parts to placing a microphone, one is in the room, the other is the distance and angle to the person speaking.

For podcasting, you may not have much choice about room placement. At least two feet or more from a flat wall is preferred. If you do have a flat wall behind your microphone, I would suggest either "treating" the wall with acoustic foam available online or at any pro audio store or getting a good microphone boundary shield.

You can buy boundary shields or make your own pretty easily. The difference comes down to budget and aesthetics. The store bought shields don't work much better than a square of acoustic foam glued to foam board or cardboard, but generally they look much nicer.

For most microphones, you want to position yourself between 8 inches and 12 inches from the microphone. If your pop filter is 3 inches or less from the mic, then a "shaka" from the pop filter to your mouth works well in most cases.

Start with that and then it will come down to trial and error for finding the best sound for your voice when using your mic in your room.

The next part of using a mic correctly is the aim of the mic. A shotgun should point right at your mouth with about a 20 to 30 degree angle. Cardioid mics like I recommend you use should face your mouth straight on.

A Cardiod should never
point at you. Notice how
the shaka hand points at
the mouth and the face
of the microphone.
When an Omni or
Shotgun is put into a
stand it will look like it is
pointing at you. A cardioid should not.

Wrong!

Pro Tip -

USB microphones have all of the electronics built in. As you record more podcasts you might want to upgrade to XLR mics so that you don't have one component that puts you out of business, or is hard to find last minute.

8. Record It.

Now it is time to put your voice and system to the test. I have to warn you, if you aren't used to hearing yourself recorded, you probably won't like what you hear. I have seen many people try and record and say "I hate my voice" and quit.

Let other people hear it. They hear you all the time, it is just different to you. Remember the first time you heard yourself on voicemail? It isn't that bad.

The first step is to set up your recording. Since you are only recording one microphone and one voice, you only need Mono. Stereo can do funny things. **** 44khz 16bit is pretty normal for Voice Recording. If you are sending something in, always ask. For your Podcast, 44khz/16bit/mono will work perfectly.

Setting up your microphone is the next step. Earlier we talked about how to find a good distance. Now it is time to test it out and set the gain. The gain is like the volume on the mic. The higher the gain the louder you are when you record. If you are two quiet, the room noise can get in the way, and if you are too loud you might cause distortion. There is a fine balance.

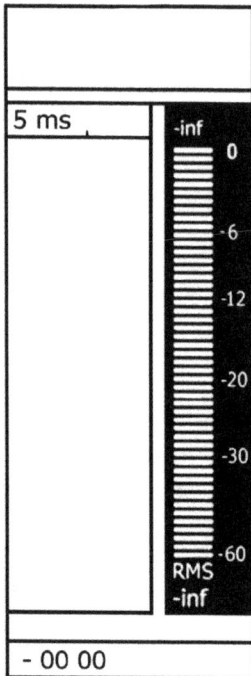

```
 ┌─────────────────────┐
 │                     │
 │                     │
 ├──────────┬──────────┤
 │ 5 ms ,   │  -inf    │
 │          │ ▤  0     │  Red To Much Gain
 │          │ ▤        │       or
 │          │ ▤ -6     │  To Close To Mic
 │          │ ▤        │
 │          │ ▤ -12    │  Yellow Peaks
 │          │ ▤        │      Good
 │          │ ▤ -20    │
 │          │ ▤        │  Top of Green
 │          │ ▤ -30    │      Best
 │          │ ▤        │
 │          │ ▤ -60    │  Bottom of Green
 │          │ RMS      │     Too Low
 │          │ -inf     │
 ├──────────┴──────────┤
 │ - 00 00             │
 └─────────────────────┘
```

When I speak to the mic I like to have my peaks at about -3 or -4 db on the meter. For most software packages, that is in the high green to low yellow band. Over time you will learn if you get excited or loud, you need to pull away from

58

the mic a bit, and if you get quiet for effect, move closer.

After your mic is set, you are ready to record, and the first thing to do is record about 10 seconds of silence in your room. Try not to even breathe for the 10 seconds, just so you get the pure acoustic signature of the room.

Play it back and listen to everything you hear. Ideally your noise floor should be about -54 db on the sound meter. Some software allows you to "monitor audio" without recording. This is good for determining the noise floor and maybe setting the mic gain, but you will only know what the room sounds like by recording it and listening to it with good headphones.

Pro Tip -

The aim and placement of the microphone directly effects the recording level and quality. If you are going to have a loud passage, pull back from the microphone.

Turning sideways doesn't always work, and in fact if you turn sideways and lean slightly back you could actually increase the sound picked up by the microphone. This is a common mistake.

9. Clean Up Your Work

If you want to sound like a pro, you need to clean up your work. If you don't have a director or an editor, you will need to be your worst critic. The toughest part is balance.

Perfection is the enemy of greatness.

You can spend too much time making it perfect. A basic clean up of your first few podcasts will be fine. I personally don't use the "replace with silence" feature because it can be too quiet and sound like your podcast ended. At the very least it sounds edited.

Even though you have a script and are going to edit the final product, your goal is to create something natural sounding. You want your fans to listen and think you are right there in their head talking to them and no one else.

You also need to sound professional, no "ums" or "ahs" or "yea…yea" fillers. You want to sound nearly perfect, not perfect.

The way I like to clean up my podcasts is to record about 10 seconds of silence at the very end of my recording. Engineers call this "capturing the room". When you do this you won't see a perfectly flat line, every room has a little energy that the mic can hear even if you don't. That energy is what makes your podcast sound natural, especially when your fans are wearing earbuds or noise canceling headsets.

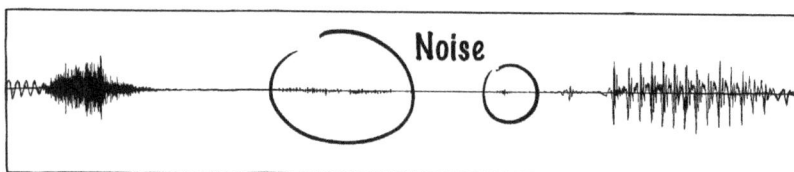

After you capture the room, copy about three seconds worth of clean room sound. Every time you smack your lips, breath too loud, sniffle or cough, simply just paste the two seconds over the "noise".

Another trick is to reduce the gaps in your sentences. People listening are different that people watching on stage. Silence over one second and they start to think about something else.

Finally you want to save your file in a format that iTunes will recognize and accept. M4A, MP3, MOV, MP4 are all acceptable formats.

Pro Tip -

Cleaning up and editing your work is the icing on the cake. Even if the cake is ugly or doesn't taste that good, doesn't the frosting always make it look good?

Sound isn't any different. Get rid of the "Uhms" pops and "ahs" and breathing and the rest might turn out just fine.

Don't start over until you edit at least a little and see what you got. If the trash truck pulls in, then start over for sure.

10. Get Your Work Online

Getting your podcast online is different than getting a song online. You can upload music to the apple store and they store it, advertise it and send you money if enough people buy it. For the service Apple will take a cut.

The same is true for books and sometimes videos. Podcasts are different because you can't charge for them, one requirement of a podcast is that it has to be free in order for Apple to catalog it. Because Apple doesn't make any money, you have to "host" your own podcasts online. Starting out, just about any website hosting provider that you pay for will include enough storage and bandwidth for 100 or so podcasts for less than $20 a month including your website.

WordPress.com will let you build a site for free if you just want to try it out. Most "safe mode" or free installations of Wordpress will be limited to a 2MB file which isn't much. You will eventually want the hosted version, not in "safe mode" if you are going to use WordPress as your web site engine.

If you need more than that, Amazon and other companies offer very cheap cloud storage services where you can have hundreds of people download your podcast for a few pennies. If you start making videos or selling video content this is a little better than YouTube because you don't have to worry about competitors ads getting on your video.

You can of course pay YouTube or monetize it which is a nice side benefit. Also if you get enough subscribers and live in the Los Angeles area, you can use the YouTube Studios for free. Video can have some huge benefits.

To move your recorded podcast online, you have two pretty easy choices. One is to use a Wordpress™ website either on wordpress.com or hosted on your on internet service provider or web hosting provider.

If you don't use Wordpress, you will likely need some sort of FTP program. Windows has basic ftp built in to the file manager. Apple sort of does, but spending $5 on a good app is well worth the money. I use Yummy FTP Lite, it works great and is easy to use. When you start making a little money you can hire someone else to help you, or at least spend a few dollars and buy the full version of an FTP program.

What is FTP?

If you don't know how to use FTP or what it is, here is a basic introduction. FTP means *File Transfer Protocol*. Simply put it is the easiest way for you to move a file on your computer to a file on the internet, or transfer a file from your computer to theirs.

It isn't quite as easy as moving a file from one folder to another the first time, but once you set up your FTP program that is about it.

You will need your ftp username, password and path.

Many internet providers will give you a slightly different username for using FTP. This is for better security.

The ftp path is the location of the webserver where you will be moving your files. Many hosting companies simply use ftp instead of www

on your domain. So your ftp path would be
ftp.yourwebsite.com where "yourwebsite" is the
name of your website.

Your internet hosting provider like *1and1.com* or
GoDaddy will provide you with this information.

Basically your ftp software will look like this before
you connect. The folder on the left is on my
computer, and the folder on the right is on my
website.

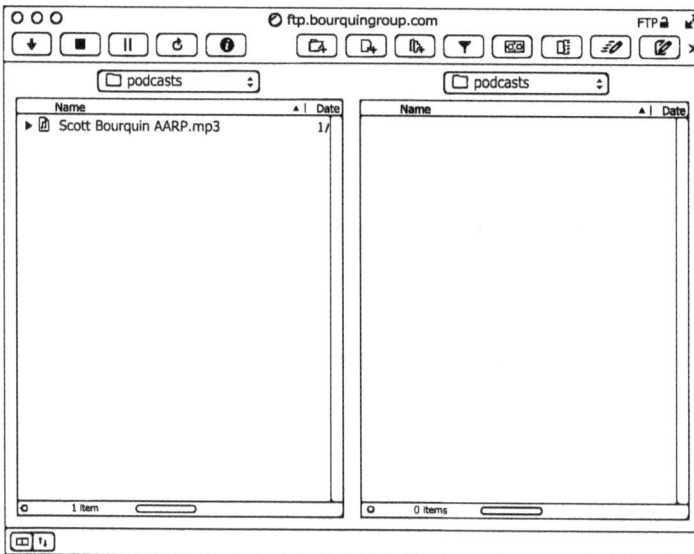

Once you set up your FTP software and connect to the server, you should have two windows or two columns in one window. One with the files on your computer, and one with the files on the server. With most FTP software you just click on the podcast to highlight it, click the little arrow to transfer it to the server.

The only real trick is finding the right folder to move it to.

The Wordpress Method

If you have a Wordpress™ based website, there are podcast hosting plug-ins that can make this process easier. I have found a few that make it harder though, so no real recommendations here, pick one, try it and if it works, keep using it.

This is exactly what the page looked like on my website when I posted my first podcast. the link takes you to a page where you can listen.

You can see the page at www.scottbourquin.com/podcasts. You can also see a direct post podcasts at http://www.bourquingroup.com/podcast.

If you want to update your podcasts on iTunes the other requirement is that the site has an RSS feed that is RSS 2.0 compliant or better. RSS means "really simple syndication" and RSS is how your podcast will be "syndicated" so it can be shared by more people and others can be

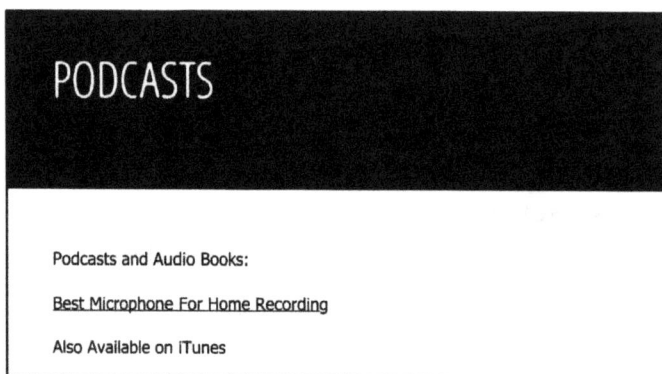

PODCASTS

Podcasts and Audio Books:

Best Microphone For Home Recording

Also Available on iTunes

automatically notified when you create a new podcast. If you have a Wordpress website, there is a plug in for this too.

Can you tell I like Wordpress? A quick side note about Wordpress™. Many computer and web programmers don't like Wordpress™. The simple reason is they can't control it. After they build you a website, you don't really need them anymore. With other Content Management Systems or CMS based websites, you need your programmer for every little change.

You can get a free account, upload a picture and build a profile at www.wordpress.org, and you can host a free website at wordpress.com. I haven't verified that wordpress.com will allow you to store Podcasts and host RSS there however the instructions are on the FAQ page so it is likely.

http://en.support.wordpress.com/audio/podcasting/

Pro Tip -

Starting out WordPress is the way to go. Very little learning curve, and much better support.

Unless you are web programmer wait until the money is coming in and pay someone to do that for you.

11. Publishing to iTunes

Publishing to iTunes can be a bit frustrating at first. There is a very specific process and you need an iTunes account. The good news is like everything else in life, the more podcasting you do, the easier it gets.

Your don't really publish a podcast to iTunes. What you do is tell Apple where the podcast is, they review it and make sure you aren't posting illegal or inappropriate content and then list the podcast in the Apple store along with millions of other podcasts.

That is really it. You will wait up to a week to get a very anti climactic e-mail letter that says "your podcast is online".

The Steps:

1. After your podcast on your website, you need to create an RSS feed (an XML file) that:
    - Conforms to RSS 2.0
    - Includes recommended iTunes RSS tags
    - Contains pointers to your episode with the

<enclosure> tag.

2. Create your cover art, your picture is ok, or you can have something cool made fore a few bucks at fiver.com or odesk.com

3. Post the RSS Feed and cover art along with at least one episode of your podcast to a server that supports byte range requests and a url that is publicly available.

4. Get an Apple iTunes Account. You probably have one, and if you have an iPhone you do.

5. Go to the submit a podcast page at iTunes and follow the directions. They are pretty simple:

   http://www.apple.com/itunes/podcasts/specs.html#submitting

   On this page when you get to the right place is the submit a podcast link which will walk you through uploading the link to iTunes and starting the approval process. This isn't the path to your podcast, this is the URL to your Feed. Slightly different. I used feedburner.com for my feed host. So the URL I submitted was http://feeds.feedburner.com/scottbourquin

6. Wait

7. Wait

8. Apple will send an email that your podcast is approved or denied. If it is denied they may or may not tell you why. Most likely it is for a copyright violation meaning you used music or read a book and didn't include that you have permission. If it is denied - Try again.

When your podcast is up, if you search your name on iTunes you'll find something like this:

Podcast Episodes						See All >
Name	Provider	Podcast	Time	Popularity	Price
Which Microphone is Best For Home...	SJBourquin	Scott Bourquin			Free ▾
		See All			

9. Test your Podcast, find a friend to subscribe and listen to it. Have them be very harsh as a

critic. Was it easy to find, did it download easily, did they listen, and did they like it? Why or Why Not? You want to know all of this to make it better.

Pro Tip -

When you post an RSS feed to iTunes, it automatically adds new podcasts each time. It can take up to five days when you podcast is new and you don't have a ton of followers.

After you post your second podcast, go back and make sure it shows up.

Notes From Apple iTunes Podcasting –

Additionally, we recommend that you:

1   Pay close attention to the title, author, and description tags at the <channel> and <item> level of your podcast feed. The iTunes Store uses these fields for search. The metadata for your podcast, along with your cover art, is your product packaging and may affect whether your podcast shows up in relevant searches, and how likely users are to subscribe to it.

2   Make your title specific. A podcast called "Our Community Bulletin" is too vague to attract many subscribers, no matter how compelling the content.

3   Take advantage of the <itunes:summary> tag. The <itunes:summary> tag (or the <description> tag if <itunes:summary> is not present) allows you to inform users about your podcast. Describe your subject matter, media format, episode schedule, and other relevant information. In addition, make a list of the most relevant search terms for your podcast and build them into your description. Note that the iTunes Store removes podcasts that include irrelevant words in the <itunes:summary> or <description> tags.

4   Be sure to include a valid
     <itunes:category> tag. You must also
     define a subcategory if one is available.
     Podcasts with category information appear
     in more places on the iTunes Store and are
     more likely to be found by users. Your
     category should be in English in the XML
     of your feed, but will be localized on the
     iTunes Store.
5   Pick a reliable webhost. Podcasters
     sometimes create a feed and then find that
     their ability to move or edit the feed later
     is limited by the webhost they have
     chosen. To avoid this, make sure your
     podcast is hosted on a website where you
     are in control.
6   Create cover art for your podcast that still
     works well when scaled down to thumbnail
     size. Before you create your podcast cover
     art, review the Top Podcasts section in the
     Podcasts app or in the iTunes Store for
     examples of compelling cover art. To be
     eligible for featured placement on the
     iTunes Store and the Podcasts app for iOS,
     cover art must be a 1400 x 1400-pixel
     JPEG or PNG file in the RGB color space.
     You can also use specific images for
     individual episodes.

7   Enclose all portions of your XML that
    contain embedded links in a CDATA
    section to prevent display issues and to
    ensure proper link functionality in the
    Podcasts app. For example:
    <itunes:summary><![CDATA[<a
    href="http://www.apple.com">Apple</
    a>]]></itunes:summary>

Marketing To Spread The Word

The last step is to market your podcast. I realize this sounds counter intuitive to spend time and money to give something away, but you didn't record it so no one would here it did you?

If you've taken the time to got through steps 1 through 7 and share a little bit from within you, share it, market it and be heard. Someone wants to hear you, and more importantly someone might need to hear you.

Ideally as you add new podcasts to your feeds, iTunes will Automatically update. If not double check your settings at your host to make sure iTunes can see them.

12. Marketing.

Recording a Podcast and getting it online is an adventure all by itself sometimes, but if no one hears it, what was the purpose? You can record yourself and listen to yourself all day long and not have to deal with FTP, Apple or even a website.

The whole point is to record something that people want to hear. Something that tells them a story or teaches them something they want to know.

Until you are Rush Limbaugh, spouting out rants won't get you far. If you do it comedically, you might have a chance. The better way is to tell a story and hide the lesson in the story. At the end of the day your podcast needs to have some level of entertainment value or people won't listen. The truth is no one cares about your story or your lessons unless they can relate and make it their story. Make it all about them. Answer questions they have.

Just as important as recording the podcast is marketing it, letting people know it exists. Just like creating an "app", Apple lists hundreds of thousands of podcasts, and yours starts out as a very small needle in a pile of different needles all covered by a mound of hay.

Marketing your content is how you help people find your podcast.

Your List

One thing you will hear in internet marketing circles over and over again is the term "the list" or "your list".

Your list is nothing more than a list of people who are willing to let you send them content. Most of the time the content is simply an email, other times it is a book, sometimes it might be a podcast. There is a great course on list building if you are building a virtual business or online business by Jeff Walker, called PLF in the industry. You will see Jeff's name in just about every online marketing book there is.

If you are on our newsletter list, the next time Jeff has a class I will share it with you. You can join that list at www.easyguidebooks.com/newsletter

If you really want to find out more about how to market your podcasts better, check out the **Easy Guide To Marketing**. Just paying for ads is like throwing money away. Knowing why ads stick and creating sticky ads is a license to print money.

Pro Tip -

The key to a successful podcast is to offer information that really helps people. Something they want to hear. So before you even start, always ask, do they care about this or do I?

If you care about it more than your audience it can should like a rant or get preachy. Rush Limbaugh might have built an audience that way, but it isn't the Easy Way.

Once your podcast is online and you have started marketing it, it is time to go back to step one and do the next podcast. The more material you have the more likely you are to build a fan base.

The good news is that if you set up your RSS feed correctly, you don't need to tell Apple or post to iTunes. The RSS feed will do that for you automatically!

What Story Are You Going To Tell?

The world is listening!

Scott

www.ingramcontent.com/pod-product-compliance
Lightning Source LLC
Chambersburg PA
CBHW060640210326
41520CB00010B/1677